Ali Raheem

An Investigaton into Authetication security of GSM algorithm for Mobile Banking

Anchor Compact

Raheem, Ali: An Investigaton into Authetication security of GSM algorithm for Mobile Banking. Hamburg, Anchor Academic Publishing 2013
Original title of the thesis: Development and Implementation of secure GSM algorithm for Mobile Banking

Buch-ISBN: 978-3-95489-077-4
PDF-eBook-ISBN: 978-3-95489-577-9
Druck/Herstellung: Anchor Academic Publishing, Hamburg, 2013
Additionally: Middlesex University in London, London, England, Bachelor Thesis, 2011

Bibliografische Information der Deutschen Nationalbibliothek:
Die Deutsche Nationalbibliothek verzeichnet diese Publikation in der Deutschen Nationalbibliografie; detaillierte bibliografische Daten sind im Internet über http://dnb.d-nb.de abrufbar

Bibliographical Information of the German National Library:
The German National Library lists this publication in the German National Bibliography. Detailed bibliographic data can be found at: http://dnb.d-nb.de

© Anchor Academic Publishing, ein Imprint der Diplomica® Verlag GmbH
http://www.diplom.de, Hamburg 2013
Printed in Germany

Abstract:

GSM systems are vulnerable to an unauthorized access and eaves droppings when compared with the traditional fixed wired networks due to the mobility of its users. The main idea of this project is to develop an application device to secure mobile banking over unsecure GSM network. It is important to mention here that, mobile banking is a term used for performing balance checks, account transactions, payments, credit applications and many other online applications. But unfortunately, the security architecture for cellular network is not entirely secure. As a matter of fact, GSM network infrastructure is proved to be insecure and many possible attacks have well documented in the literature. Security was never considered in the initial stages therefore sending protective banking information across an open mobile phone network remains insecure. Consequently, this project focuses entirely on the developing and designing security techniques to asses some security issues within mobile banking through cellular phone network (GSM). The main aim of this project was to investigate and examine the following:

1. Security issues in each level of the mobile network architecture.
2. Messages and signals exchanged between user's cellular phone and mobile network at each level.
3. The overall security architecture of GSM flaws.
4. Some existing security measures for mobile transactions.
5. The current security within SMS banking and GPRS banking.

Finally, two folded simulation in MATLABT were performed using OFDM which is a broadband multicarrier modulation method that provides a high performance operation to transmitted and received data or information. In other words, it is the most customary single that uses carrier modulation that gives high speed function in microwave frequency. Therefore, the first program was concerned with generating transmission and receiving the OFDM signal without channel noise effect. The second program was concerned with the effects of high power amplifier and channel noise on the OFDM signals. It is to be noticed here that the OFDM is a modulation that is especially suitable for wireless communication. Consequently, the suggested programme succeeded in achieving a limited noise or interference in the signal as the users complained and suffered constantly from this noise and from losing the data or the information.

1

Table of Contents:

List of Figures:

List of Tables

Abbreviations

1.GSM	Global System for mobile communication	37.RAND	Random number
2.GMSK	Gaussian Minimum Shift Keying	38.Ki	Key infrastructure
3.TDMA	Time Derision Multiple Access	39.IMEI	International Mobile Equipment Identity
4.LBS	Location based Services	40.DOS	Denial Of Service
5.MATLAB	Mdrix Laboratory	41.USSD	Unstmctured Supplementary Services Data
6.OFDM	Orthogonal Frequency Division Multiplexing	42.PIN	Personal Identification Number
7.SMS	Short Message Service	43.WML	Wireless Markup Language
8.M-baking	Mobile baking	44.TLS	Transport Layer Security
9.GPRS	General Packet Radio Service	45.SSL	Secure Sockets Layer
10.FDMA	Frequency Derision Multiple Access	46.WTLS	Wireless Transport Layer Security
11.SGSN	Serving GPRS Support Node	47.APN	Access Point Name
12.IMSI	Subscriber identity	48.SGP	Service General Packet
13.MS	Mobile Station	49. AES	Advanced Encryption Standard
14.BTS	Base Transceiver Station	50. SHA-1	Secret Key Hash
15.MSC	Mobile Switching Centre	51.RSA	Rivest Shamir Adleman (algorithm for Public Key cryptography)
16.HLR	Home Location Register	62. ID	Identification
17.VLR	Visitor Locator register	63.4G	Fourth Generation
18.SMSC	Short Message Service Centre	64. SS7	Signalling System No7
19.SIM	Subscriber's Identity		
20.WAP	Wireless Application Protocol		
21.ISMG	Internet Short Message Gateway		
22.SMPP	Short Message Peer to Peer Protocol		
23.2G	Second Generation		
24.3G	Third Generation		
25.PKI	Public Key Infrastructure		
26.WIG	Wireless Internet Getaway		
27.NBO	National Bank in OMAN		
28.BSC	Base Station Controller		
29.OMC	Operation and management centre		
30.SMSC	Short Message Service Centre		
31.ISC	International Switching centre		
32.AUC	Authentication Centre		
33.HLR	Home Location Registry		
34.PDP	Packet Data Protocol		
35.GTP	Tunneling Protocol		
36.GSN	Support Nodes		

Acknowledgment

I would like here to express my sincere thanks and deepest gratitude to my supervisor Dr. Abou-baker Lasebase, Head of Telecommunication Department. His constant guidance, painstaking reading and valuable instructions hare helped much in writing this research. His course on Telecommunication Security has helped a lot in developing the topic of this research.

Thanks are also extended to all the staff members who taught and supported me though out my whole study.

I would like also to express my gratefulness to my father, Professor Dr. Hussein Raheem and my mother professor Dr. Ibtisam Jasim, who have shared with me all my moments of aspiration and desperation. Their financial and moral generosity is beyond expression. I would like also to thank my sister and colleague Rand whose kind presence with me helps a lot in facing all the difficulties and challenges as an international student

My last but not least thanks are for all my friends and for all those who have helped me in one way or another in the different stages of my study.

Chapter 1: Overview of mobile banking security

1. Introduction:

We present age is witnessing rapid and continuous technological inventions. GSM is one of the greatest derives that man has ever invented and which is at present prevailing our planet. Today almost everyone uses GSM system in one way or another. The fact that infrastructure is already there makes this device easy and convenient to be used at the different types of human communication. In reality, GSM communication channel call is only partly encrypted. The security and authentication mechanisms incorporated in GSM make it the most secure mobile communication standard currently available, particularly in comparison with the analogue systems to be described. Part of the enhanced security of GSM is due to the fact that it is a digital system utilizing a speech coding algorithm, GMSK, digital modulation, slow frequency hopping and TDMA time slot architecture. In addition, GSM high security standard makes it the most secure cellular telecommunication currently available. Although the confidentiality of a call and anonymity of the GSM subscriber is only guaranteed on the radio channel, it is a prerequisite in achieving end to end security. The subscriber's anonymity is ensured through the use of temporary identification number. The confidentiality of the communication itself on the radio link is performed by the application of encryption algorithms and frequency hopping which could only be realized by using digital system and signalling. Mobile Banking refers to the provision and availability of banking and financial services with the help of mobile telecommunication devices. The scope of offered services may include facilities to conduct bank and stock market transactions, to administer accounts and to access customized information. With the rapid development of technology, the mobile telecommunication technology has widely expanded in different world countries. Mobile banking becomes especially important and popular these days .It can provide customers with faster and highly qualified services to perform their banking and financial services with the assistance of mobile telecommunication devices. In addition, the majority of the mobile banking researchers have agreed that mobile banking consists of three main sections, namely mobile accounting, mobile brokerage and mobile financial which contains the information necessary for such services. On the other hand, the customer service sector has many processes such as balance checking, account transactions and payment. Furthermore, mobile banking can provide different facilities, such as LBS affiliated to the banks. Contrasted with the Internet banking, the mobile banking is completely secure and easier to be used. Added to this, mobile banking does not only provide basic services like traditional banks, but it can provide the customer with 3A services simultaneously; these services are anytime, anywhere and anyhow. Unfortunately, nothing is perfect in life, the security architecture for cellular network is not entirely secure because the GSM network infrastructure has been proved to be insecure and many possible attacks have been discovered as protection discretions are not well considered. As a result, sending protective banking information across open mobile phone network is entirely insecure. Accordingly, Margrave D (2006) [9], attempted to cover the security issues within mobile banking through cellular phone network (GSM). The goal was build portable device applications that ensure that users can send securely their banking information via mobile network. Obviously, the rapid growth of digital wireless communication in recent years has increased the need for high speed mobile data transmission. Now, modulation techniques were implemented to enhance communication capacity. Simulation by using MATLAB has been implemented to generate, transmit and receive OFDM signal without a channel noise effect and also to examine the effects of high power amplifier and channel noise on OFDM signals. The project, therefore, aspired to develop an application device to secure mobile banking over unsecure GSM network. Consequently, the project will deal also with some network security requirements such as authentication, non repudiation, authorization, availability, integrity, confidentiality and access control depending on the insecurity of existing protocol. However, it is quite possible that expected attacks were produced. Significantly, these attacks unravelled flaws existing in mobile banking protocols. The project, therefore, had algorithm for Mobile Banking to provide a convenient and effective means for customers to pay and perform other banking transactions. To sum it up , mobile banking service is a modified version of internet banking using cellular technology and GSM network as a medium to transfer request over wireless network as portrayed in the following picture.

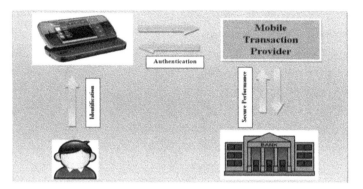

Figure 1: Modular Transaction Architecture

Chapter 2: GSM Security Issues

2. Literature review:

The aim of this part is to summarize and asses some research writings that have attempted to approach the topic under questioning. It intends also to show any possible connections and relevancies of importance between them and the subject of this research. This review is carried out through the discussion of four important sections:

- GSM network architecture and Security.
- Mobile Banking.
- SMS/ GPRS Banking Services
- End to End security architecture for the mobile banking system

2.1 GSM network architecture and Security:

The mobile wireless networks are somehow susceptible and vulnerable to unauthorized users access. This is because of the mobility and the flexibility provided by this GSM network to all users regardless of the varied nature of the covered area or place. GSM can be regarded as a system of mobile communication. In New Generation Features for GSM Systems, Liveris, Anagnostostopulos, Lykou and Stasinopoulos (2010) [10] pointed out that GSM offers enhanced features such as total mobility, in the sense that the subscriber has the ability to communicate in different areas covered by the GSM cellular network outside his home location or even out of his country. It can also provide an excellent and high capacity achieved by using the FDMA, TDMA, efficient speech coding and the GMSK modulation scheme. Furthermore, it introduces different services to the users such as voice communication, voice mail, short message transmission etc. Finally, they concluded that GSM possesses the most secure cellular communication standard. They enumerate the GSM features, but they did not tackle or discuss the vulnerabilities in the GSM network architecture.

Kasim and Eraul (2010) [1] similarly discussed and evaluated the security techniques in the GSM. However; they attempted to set a new protocol that can improve the security of the GSM. Regarding GSM architecture, they stated that GSM network can be divided into three important areas. The first area is the MS which is simultaneously a base for both the station of the subsystem and the network of the subsystem. Furthermore, the MS consists of two main important elements: the mobile equipment and the subscriber identity module. Moreover, the BSS controls the radio link and connects it with the MS. Besides, the network subsystem provides and performs the switching of the calls among mobile users and mobile fixed networks such as ISDN, PST...etc. Accordingly, Figure 2 shows the distribution of GSM security parameters and also the IMSI. When the cellular phone user moves from one domain to another, the user's identity should be recognized at every domain or area of boundary encountered. The IMSI remains confidential, as it provides the user with privacy .The identity of the users or the subscribers are protected from any hacker who might cause damage to the network system. Those researchers went on to say in their article that making authentication to GSM can provide mobile communication system with more security. Consequently most GSMs are based on the authentication and encryption of the information (data) techniques which are capable of preventing hacker from stealing the information of the users. On the other hand, these techniques are not totally void of weaknesses or vulnerabilities. For instance, confidentiality is guaranteed only when the radio channel is exclusively between the MS and the BSS. Another problem arises when the mobile moves from one domain to another. Obviously, such movement has its direct effect on the quality of the security services. Furthermore, the user's identity confidentiality depends on the user's information in accessing the net work. In other words, when the user provides the system with wrong data, the system will break down. Therefore, this article gave indeed illuminating information security in GSM, but it does not mention how this new protocol could avoid different attacks, such as the Dos and man in the middle attack.

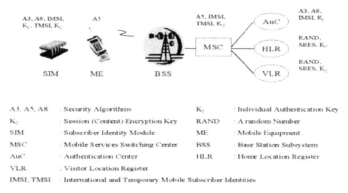

A3, A5, A8 : Security Algorithms K₁ : Individual Authentication Key
K_c : Session (Content) Encryption Key RAND : A random Number
SIM : Subscriber Identity Module ME : Mobile Equipment
MSC : Mobile Services Switching Center BSS : Base Station Subsystem
AuC : Authentication Center HLR : Home Location Register
VLR : Visitor Location Register
IMSI, TMSI : International and Temporary Mobile Subscriber Identities

Figure 2: Distribution of GSM Security Permanents [1]

"In solutions to the GSM security weaknesses", Toorani and Shirzi 2010 [1] tackled some significant security vulnerabilities and flaws and attempted some practical solutions to improve the security of the currently available 2G system. In addition, they discussed some vulnerability in GSM System, such as Man in the Middle Attack, lack of user visibility, vulnerability to Denial of service Dos attack and vulnerability to replay attacks. As solutions, they suggested, for instance, using the secure algorithms for A3/A8 implementations that can protect the SIM card from any cloning attack. This suggestion was quiet reasonable because the network operators could improve themselves without any aid from software and hardware manufacturers or GSM consortium. Added to this, they offered another solution namely using the ciphering algorithms where the operators could use secure algorithms, like the A5/3 which can be implemented on both the BTS and the mobile phones. Any change in these algorithms would prevent communication between the BTS and the mobile phones. The idea behind this solution was to encrypt the traffic between the network components so as to prevent the attacker from modifying the transmitted data. The researchers, however, recommended highly using the End to End security for being the best, the easiest and the most advantageous solution. The significance of Toorani and Shirzi's paper lies in the fact that it is somehow related to the topic of this project especially in matter of developing an algorithm which can improve the mobile baking security.

2.2 Mobile Banking and security:

Definitely, the mobile banking has become greatly important nowadays, because it is faster and easier to be used and it saves time. However, security is a big challenge to this system because hackers form the most dangerous threat to this system. They can steal data and damage the system. Such issues were discussed by different writers in an attempt to identify security problems in banking application and to suggest remedial means to these problems. To begin with, Amir Herzberg (2003) [2] has discussed some challenging issues related directly to secure payments in the banking transaction process. He provides a modular architecture which supports the security when the transactions start from the bank to the users. In addition, his architecture consisted of three important independent processes. In the first one, the device could identify the user from his card, password or from the information that was stored in the database of the system. The second process was authentication which was considered as the most important process because it could identify the user on the network of the mobile banking. Also, it supported the security in this system; the mobile authenticates the transaction request through either subscriber identification or through the cryptographic mechanism labelled as public or private key. The third process was the secure performance, which depended on the by the mobile transaction quality provided to the user. Nonetheless, these processes do not provide good secure implementation to the mobile banking transaction; therefore they cannot be taken as reliable and trustworthy.

According to Chou et al. (2010) [12] the mobile banking payment architecture connected up the provided operator service provides and banking institution. He went on to say that the architecture permitted mobile consumers to buy services using the SMS and WAP. Therefore this architecture just provides an alternative payment idea but does not offer a clear picture concerning the process service delivery.

Elliott et al (2010) [14], however, established a payment system which used brands restrictive blinding signature to the mobile devices which could offer a multi-party security .As for Buchana et al [6], he combined the SET protocol and the TLS/WTLS protocol to enhance security services over the WAP 1.X for the payment in the m-commerce. Kungisdan et al. [9] on the other hand, suggested a payment protocol which provided a good measure of security. It is to be admitted here that this protocol meets the transaction security pre requisites using the public key based on payment protocols like the SET and the IKP.

Laforet (2006) [27] thought that mobile Internet banking is slowly increasing in China .At that year, there was only 33 % using online banking and 14% using mobile banking. Whereas, Howecroft et, (2005) [28] reported that younger consumers used mobile banking more than older consumers. Skmm (2005) [30] considered mobile phone as a good communication tool for the users to do their banking work in short time and from anywhere and at any time.

2.3 SMS/ GPRS Banking Services:

Since 1991, text messaging has been developing into GSM digital phone rapidly. According to Baron, Patterson, and Harries, (2010) [37] SMS short service is very crucial for mobile banking services. Harb and Farahat (2008) [19] believe that the SMS was the main medium for mobile banking for being quick and easy to be used by customers. A recent survey performed by Hudson (2008) [17] using SMS also emphasizes the dramatic increase in the last few decades, and as an example he mentions that the New Zealanders send 600 million messages per month. The number stated showed that the SMS become a very familiar device in mobile banking. The revolutionary development of the information technology enhanced the electronic banking process. According to the Zhang (2010) [48] the SMS mobile Banking is a new model that uses the SIM card together with the STK technology .The latter service is available in mobile phones via using different networks such as GSM, GPRS and CDMA. These techniques do not only save information in one place but have also a high process capacity represented in the micro- processing unit (CPU). Added to this, the STK is completely different from the SIM card which is used only for the user's identity recognition. Furthermore, the information communication between the customer and the SMS mobile banking business centre can be completed only with a third party. Here the mobile telecommunication company makes a connection via using the GSM or the GPRS networks, using the internet short message and SMS mobile bank business centre. Therefore, the short message gateway provides a good exchange channel for the short message centre and the mobile bank business. In other words, the short message centre uses also the short message peer to peer protocol (SMPP), which subsequently makes communication with internet short gateway. On the other hand, SMS mobile bank can be attacked by virus attacker or denial service attacker, which can damage and loose the information. Although, this technology is very cheap and fast, it is very fragile in the face of attackers. In addition, the writer does not offer any positive solutions to the above mentioned problems. Similarly, Tekelec (2007) [40] also emphasizes the significance of the SMS banking services. But, he confirms too that this service was widely exposed to the unauthorized SS7 network hackers, who can spoof   the data and use it for their own interests. GPRS used both 2G and 3G. However, GPRS 2.5 G   had the ability to access the internet and also use multimedia messaging services such as the WAP. According to M.Shirali (2010) [25] the GPRS was one of greatest technologies that had been ever invented. The 2G GPRS service used data from 56 to 11 Kbit /second in the dial up modem speeds, which was rather slow. The 3G of the GPRS wireless network was indeed much faster. Paeng (2000) [1] tackles both the GSM and the GPRS security and authentication. The GPRS was based on the A3 algorithm, but he does not mention the security problem in A3 algorithm GPRS and he did not suggest any solutions which could change the performance of authentication.

2.4 End to End security architecture for mobile banking system:

Narendiran, Alblbert, Rabara and Rajendran (2010) [13] attempted an evaluation of End to End security architecture for mobile banking system. They proposed End to End security framework by using KI for the mobile banking. They pointed out how to encrypt the messages by using the public keys .End to End security performance guarantees the data transmitted using the X.409 sander. This article provides a lot of information about this technique, but it used an old version of sander certificate. We recommend, therefore, using a new sander such as X500, X509 and SSL. The problem with End to End security in mobile banking is that there is no full encryption data between the user and the bank, a deficiency which allows hackers to steal or destroy data.

Chapter 3: The focus of study

3.1 Theoretical section:

In the last few decades, the number of online banking users has increased with remarkable speed and cadence. This has given momentous to developers to devise more suitable methods of transactions to enable customers perform banking transactions from anywhere and at anytime. Although mobile banking is new, it is a very convenient system for users to make their transactions. The number of the bank users is being increased in a way similar to that of the mobile phone customers. The National Bank of Oman is a vivid example of this new trend in mobile banking implementation via applying the two important channels, the WAP on the GPRS, the SMS and the WIG. However, this implementation is not entirely void of weaknesses and blemishes. Consequently, this part of the project investigates the security threats faced by mobile banking implementations via using GSM networks. The aim is to build up transportable application devices so as to provide the users with more security when sending their bank information. These devices will promote the performance of the mobile banking users via two channels namely the SMS and the GPRS. To achieve this target, different points will be discussed. To begin with, an overview of the GSM architecture and the security weaknesses of this architecture will be provided .Then, the mobile banking solutions which are provided by NBO and their security shortfalls will be similarly discussed. Furthermore, suggested solutions to the SMS/GPRS banking protocol will be presented. Finally, prevalent mobile banking solutions will be compared to the suggested solutions. A comparison table will be drawn to illustrate this comparison.

3.1.1 GSM and GPRS security architecture:

Needless to say, GSM is widely distinguished in different parts of the world as a very appropriate form for mobile phone. Figure 3 shows the simple structure of the GSM architecture system that provides SMS and GPRS services.

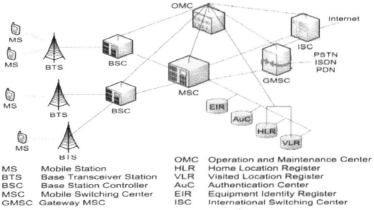

MS	Mobile Station	OMC	Operation and Maintenance Center
BTS	Base Transceiver Station	HLR	Home Location Register
BSC	Base Station Controller	VLR	Visited Location Register
MSC	Mobile Switching Center	AuC	Authentication Center
GMSC	Gateway MSC	EIR	Equipment Identity Register
		ISC	International Switching Center

Figure 3: GSM Architecture [8]

The GPRS core network can be regarded as an incorporated part of the GSM network, with added nodes to cater for packet switching. In addition, the GPRS can use some of the prevailing GSM network elements. These elements include the BSS, the MSC, the AUC, and finally, the HLR. On the other hand, a reasonable number of the GPRS network elements that can be added to GSM network consist of the GSN, the GTP and the AP, together with the PDP context. [18]

3.1.2 Security mechanisms in GSM network

Every system has some kind of security devices to vaccinate itself against attackers who aim at damaging the system. The GSM system is not an exceptional .It has some security mechanism that can hinder any illegal activities practiced by hacker's. The SIM cloning is a security device that can stop unauthorized users to access the network. The GSM has special means to authenticate and encrypt data exchanged in the network. [35]

3.1.2.1 GSM Authentication Centre

GSM authentication centre is used to provide authenticity for each SIM card available with the users. This technique provides a connection to GSM mobile network. In addition, this service prevails when the mobile connects to the network and also when the users switch on the mobile. On the other hand, if the authentication in the SIM card fails, no services network whatsoever will be offered to the user. Otherwise, if the authentication is achieved, the SGSN and HLR will allow the user to be connected to the network. [10]

3.1.2.2 Authentication Procedure

Most of the authentication of the SIM card depends on the shared symmetric secret key, which is between the card and the AUC called Ki. In addition this secret key is embedded into the SIM card through the manufacture, and is then replicated into the AUC. This authentication proceeds through several steps described as follows: [23]

1. When the AUC authenticates the SIM, it will generate a random number which is known as the RAND number.
2. The RAND number will be sent then to the users. On the other hand, both of the AUC and the SIM flow into the Ki and the RAND number .The outcome is a value known as Signed RE Sponse (SRES) and authentication is achieved faultlessly when the SIM SRES matches with AUC SRES.
3. By feeding the Ki and the RAND number into the A5 algorithm, the AUC and the SIM generate a second secret key which is known as Kc .This key is used to encrypt and decrypt the net work connection.
4. When the SIM authentication is fulfilled, the HLR will ask for the mobile identity to ensure that the MS is not black listed.
5. The mobile returns the IMEI number. The latter will be forwarded to the EIR that will authorize the subscriber .Only then; the PDP context activation takes place.

3.1.3 Issus with GSM Network System

3.1.3.1 Problems with the A3/A8 authentication algorithm:

A3/A8 can be described as the mechanism that is used to authenticate a handset on a mobile phone network. However, A3 and A8 are not really encryption algorithms. They are in fact placeholders in which COMP128is a commonly used algorithms. [28]

Suspicions concerning the validity of COMP128 as a secure medium for mobile banking were raised to the utmost when Wagner and Goldberg were able to break it in few hours .After destroying it they succeeded in obtaining the Ki and performing SIM cloning once again.

3.1.3.2 A5 algorithm Problems

Eavesdropping between the MS and the BSS is quite possible to occur .This possibility makes GPRS over the GSM network quite insecure for mobile banking. To solve this problem, A5 algorithm has been implemented. Although, there are at least three versions of the A5 algorithm (A5/1, A5/2and A5/0), they are all deficient and cannot be recommended as secure for mobile banking. [43]

3.1.3.3 Attack on the RAND value

The RAND value can be attacked by (Denial of the Service) DoS when the AUC tries to achieve the authentication with the SIM card. Therefore, The RAND value, sent to the SIM card, can be modified by an intruder who can fail the authentication. [22]

3.1.4 Current Mobile Banking

In this part, the security shortfalls of the Mobile banking solutions using SMS Short message service will be investigated.

3.1.4.1 Current SMS Banking Services in Oman:

Omani banks such as The National Bank of Oman (NBO) use (WIG) Wireless Internet Gateway for the mobile baking services. The NOB applies the USSD with the SMS. The NBO will requests the user to send the USSD cord together with his PIN to the banking server. Then, the server notifies the user that the server is prepared to accept the SMS. However, this service is not quite secure as long as each user transmits his data in plain text message. Furthermore, the mobile network operation obtains full access to the banking detailed information which is sent by the user. [21]

As a matter of fact, using SMS involves many serious security problems such as text encryption, mutual authentication, text encryption and end to end security. These problems require then a thorough investigation starting from forging originator's address.

19

The short message services can be attacked by a third party, a hacker who sends fake information to the services. It is quite possible to change the originator's address field in the SMS header to another address and this in turn enables the sender to send out fake messages. As mentioned above, the text encryption is another serious problem faced by mobile banking .The plaintext is the source of this problem. It occurs during transmission between the BTS and the MS. The End to End encryption does not exist at present .Consequently; secure algorithm is required to support security performances.

The NBO provides mobile banking GPRS, which uses MTN mobile banking gateway. This service enables   bank   account holders to access the WAP sites, in a way similar to that performed without the use of online banking. [34]

3.1.4.2 Wireless Application Protocol WAP

It is well known that WAP is an open international standard for applications which uses   wireless communication. It enables access to the internet from the mobile phone. Furthermore, the mobile phones or terminals can access the internet by using wireless application protocol browsers. The WML sites will be written instead of the HTML, XML or XHTML. In other words, the WAP protocol is confided only between the user and the WAP gateway. In this way, the connection is secured by SSL or TLS. [39] This mechanism is better illustrated in Figure 4 below:

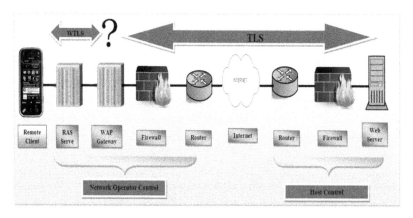

Figure 4: WAP External Network

Communication security can be provided by using the WTLS protocol and the WIM. This protocol can provide a public key based security which is similar to TLS. The WIM stores the security keys. In order to achieve interoperability of the WAP equipment and the software together with varied technologies, WAP uses the WAP protocol. Figure 5 shows the different layers of the WAP protocol. [40]

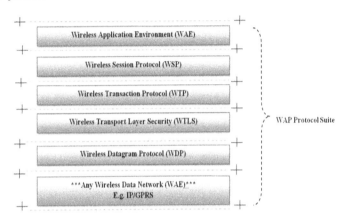

Figure 5: WAP Protocol layers

3.1.4.3 Issues in mobile banking in Oman:

The previous part was considering about how does the system of SMS messages and GPRS is working in the National Banking of Oman (NBO). However, it was quite difficult to collect the information about the security issues that might face the NBO because of the bank reputation but as an engineer it was easy to discover the security limitation in this bank. These security issues were considering about the SMS messaging protocol; therefore, a demonstration will be given and some solutions will be provided as the following shown.

3.1.4.4 Security problems with GPRS using WAP Implementations

The mobile banking which uses WAP provides a good level of security. Never the less, there are still some gaps, which could threaten the communication security. To begin with, there is no End to End encryption between the user and the bank server, but there is End to End encryption between the user and the gateway and between the gateway and the bank server. To solve this problem the bank server, should possess an APN in any GPRS networks. Significantly, this technique will be used as a WAP Gateway for the bank service .In other words, it enables the client to get connection to the bank without a third party. Furthermore, Public Key Cryptosystems provided by the WTLS are not adequate enough to meet the current WAP application security prerequisites. Because of this deficiency, the key sizes have been restricted. Last but not least, anonymous key exchange suites that are provided by the WTLT have been proven insecure. In other words, both of the client and the server are not authenticated. It is highly recommended then that banks should provide a means that could hinder it. [15]

3.1.4.5 Security problems connected with using the GPRS network

As the GPRS core network is extremely general, it needs to cater some banking security pre requisites. For example, bank authentication seems to be lacking. Without providing adequate authentication, mechanism, bank information or   client account information could possibly be endangered. Also, to avoid any fabrication of the bank's or the client's information, a provision of functions should be prevailed. Added to this, the mechanism used so far to provide confidentiality of information between the mobile station and the bank server has been proven to be vulnerable as long as the network operator can view the client's information .Moreover, specific performed actions by either the bank or the client cannot be confirmed. And this in itself is a sort of inaccuracy that might raise security matters on the bank's side's .Finally; inconsistency is another deficiency in the bank's performance that can also raise security suspicions. This problem is the outcome of the fact that the GPRS provides session handling facilities while it does not handle bank sessions. [15]

3.1.5   Secure SMS Solution

The solution for the above explained problems lies in providing a secure messaging protocol that uses SMS. This protocol has treated security problems in the current deficient GSM architecture. It is to be mentioned here that three kinds of transactions have been selected in this project, namely, check balance, transfer money and purchase airtime which can change according to the bank's services. It is also to be mentioned here that these transactions vary from one bank to another. [?]

22

3.1.5.1 Secure SMS Protocol

To make the SMS more secure for using, the SMS messages needs to be largely improved. The most important part of improving the security performance of SMS is making some changes in the message structure and protocol sequences.

3.1.5.2 Message Structure

Figure 6 shows the wireless application protocol WAP layer for the message structure. It shows the provided solution and each solution from which layer had been taken in order to solve the security issues.

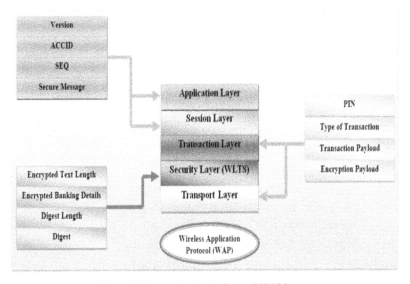

Figure 6: Wireless Application Protocol WAP Layer

To furnish the way to the different security inspection necessary for the protocol, the secured SMS message is divided into multiple fields, as illustrated in Figure 7.The numbers above the fields are the minimum number of bytes for each field, and they can be increased in accordance with usage requirements.

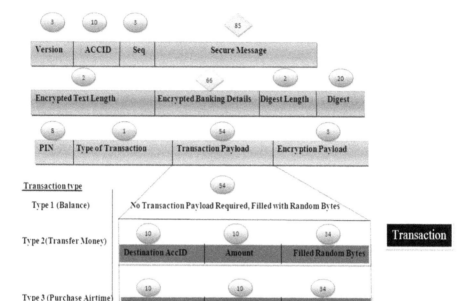

Figure 7: Structure security of SMS messages

To make it more comprehensible, the uses of all labelled structures are explained. The structure contains a specific bytes pattern. The first three bytes of the received SMS message will be checked by the receiver to prove its validity. It will be ignored if it does not match the bank application. And as accidental messages are possible, the implementation of the version bytes can eliminate them. [20]

The ACCID contains the bank account identifier of the client. The SEQ is the user's current sequence number of the one-time password. The Encrypted text length contains the number of next bytes which has the ciphered message. The digest Length contains the number of next bytes that enclose the digest. The Digest, contains calculated digest value of the message used to test the message integrity .To perform secure SMS banking protocol, a single digest of the following fields is calculated: ACCID, Version, SEQ, PIN, and type of Transaction and Transaction Payload.

However, the content of the second field is encrypted using the generated session key. The PIN contains the user's password, which is used by the receiver application to authenticate the user. [42]

Different types of transactions can be used via secure SMS message. This type is used by the bank server application to identify the type of transaction that should be performed. Finally, the Transaction Payload is in fact an additional data or information that is used for transaction, and not for any security reason. But whereas the content of the Transaction Payload depends on the type of the transaction requested, its structure depends on the type of transaction provided by the bank.

3.1.5.3 Protocol sequences

In the GSM network, SMS messages are sent asynchronously to the receiver. As a result, the secure SMS protocol is similarly asynchronous. Figure 8 offers an overview of the secure SMS protocol.

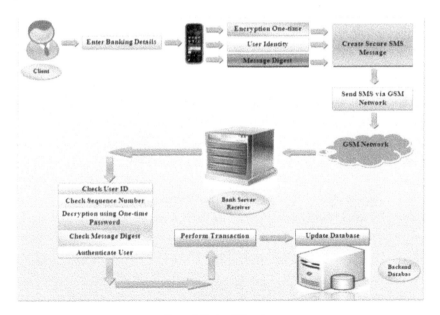

Figure 8: Protocol Overview

All mobile phones take the needed data from the users .The mobile phone generates the message and sends it to the server that reads the message and decodes the data and performs security check. Deeper understanding of the message generation and the message security checks will be offered in the following section.

3.1.6 Generating and Sending Secure SMS Messages

To begin with, the mobile phone application has a ready bytes pattern which is inserted into the message. The message will then hash value number that can check its integrity. To achieve this target, some of the content used to calculate the message digest should be encrypted by the receiver to process the message. This validation integrity will prevent attackers from generating another digest and changing the message. The protocol requires some information to identify the account holder's identity. The algorithm which is used for encryption must be a symmetric encryption algorithm. The key for encryption generates from the one time password entered by the user. It is to be mentioned here that, the pass word is only known by the server and the user. When the application processes the security contents, the SMS message will be then sent to server through the GSM phone network. When the message is received, the server will check the version bytes pattern to prove its validity for the secure SMS protocol. If the version is correct, the message is appropriate then .After that, the server checks successively the account identifier and the current sequence number. If the account identifier and the sequence number match the server database, the server will recover the one time password which can be provided from the database. The server goes on to use the retrieved password as the decryption key to decode the encrypted contents. When the decryption succeeds, the used of one time password will be ignored and server sequence counter will be incremented by value one. The server then reads the secure contents to calculate the message digest using the same algorithm used by the mobile application. To check the message authenticity, the server checks the two digests. If the message has not been altered, the server retrieves the PIN and compares it with that saved in the database. If all the security checks pass, the server performs the required transaction. [33]

3.1.6.1 Security of secure SMS Protocol

The Secure SMS protocol conforms to the general security requirements through the means of confidentiality, integrity, authentication, non-repudiation and availability: [11]

1. Confidentiality can be achieved by encrypting the message via using a symmetric secret one - time password shared only between the user and bank server. The strength of confidentiality depends on the security strength of password generation algorithm used and that of the ciphering algorithm. Otherwise, there will not be any confidentiality.

2. Integrity: As we mention in the previous part, the message digest is the hashed value of the message content calculated server application and mobile phone application. If the content has been changed through the transmission of data, the hashing algorithm will generate different digest value on the receiver side. Mismatched digests mean that the message is not secure. Therefore, providing high security depends on the strength of the algorithm encryption as well as on the digest value.

3. Authentication: To authenticate the user, the user needs to show his authentication detail to the receiver. This process is performed by validating the message PIN with the receiver PIN saved in the mobile banking account.

4. Non- Repudiation: The one time password is only held by the account holder and the bank server. In addition, the bank server cannot generate more than one time password for the sake of high security. Therefore, the onetime password is designed for a single user so that the user cannot deny not sending the message, because every user had a unique password and a sequence number to encrypt the message.

5. Availability: The cellular network is based on the availability of this protocol. The time which a message takes to be delivered depends on the quality of network operation towers. Therefore, each server has their capability to deal with a number of users.

A pseudo code for the above solution will be given as the following:

```
int (user inf)
If ( error in user inf)

{
An error message bake to user

}
else (the following process will be happened)

{

Generate 6 digits password using asymmetric key
Encrypt the 6 digit numbers
Encrypt user identity
Get message digest
Calculate the message value
Create cipher for message encryption (DES)
Send message via the GSM network
send to the Server  {
                        Check user ID
                        Check sequence number ID
                         String decrypts the one time password
                         Get the message digest decrypt
                         Store data in database

                    }

}

end
```

3.1.7 Secure GPRS Solution

To solve the security problem and produce high security performance, two important solutions are recommended which can rectify the security problem. The first is an extension of the present security features of WAP implementations and the second is a completely new GPRS security protocol which consolidates security and maximizes its performance. The suggested solutions will be explained fully below.

To begin with, extending the present WAP implementations enables bank users to connect to the bank network via a WAP gateway operating in its network system. In addition, the customized WAP gateway prohibits options like abbreviated handshake, server authenticated full handshake and anonymous key exchange suites .Figure 9 shows the perfect location of WAP gateway for mobile baking

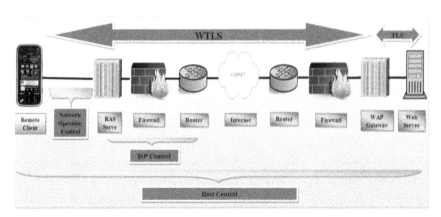

Figure 9: the Ideal WAP Gateway in banking server

The second solution labelled as the New Secure GPRS protocol has been designed not only to control security in M-commerce application but also to make and perform security connection between mobile devices and the bank servers. It is to be mentioned here that the secure GPRS protocol consists of two important components namely the initial user server and the transfer of the data record protocol using the created SGP (Service General Packet) channel and also ex-changed cipher principles suites of good privacy. [16]

3.1.7.1 Protocol message components

Every SGP message which is exchanged between the user and the server has three components namely the message timestamp, the message and finally the message type. Moreover, the message timestamp is used between the user and server to stop and limit attackers, whereas the message type is used also by the user and the server to identify the message sent. Figure 10 demonstrates the three components of the Protocol message namely error message, handshake message and go –ahead message [12]

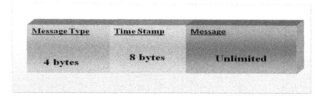

Figure 10: message components sending

3.1.7.2 Client protocol Initialization

When a client activates the mobile application, a onetime 512 RSA key duo will be generated and the client will send the public key to the server to create digital signatures that are verified by the server. This process authenticates the message that is sent by the mobile client. To complete the client protocol initialization, the client generates PBE AES key using his own password. Usually user authentication is done by the mobile device and the bank. When the user registers to use the banking service, a server certificate sign applying the user's password is included to authenticate the account holder on the mobile phone. In other words, when the user enters his password, the mobile phone application will generate the AES on the password to retrieve the server's public key available in the server certificate. Only when this process succeeds, the authentication will be achieved, otherwise, the client should enter the password again. If logging fails three times, the user account will be blocked for security purposes.

The second user authentication is performed by the server. In this case, the client sends his encrypted ID and the server will get the password from the database and then generates the AES key. The user is authenticated to access the network only when the key can successfully decode the message.

3.1.7.3   SGP (Service General Packet) handshake for Client:

The SGP handshake packs and transmits a complete SGP packet to sever to decode the message and generate the key. Figure 11 shows the packing of the full SGP packet.

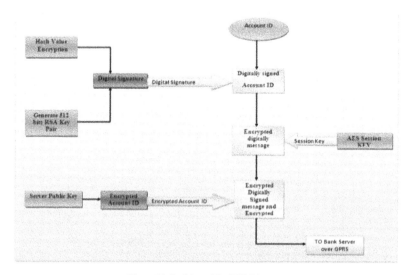

Figure 11: Packing of the SGP Message

The full SGP pack begins with hashing the client account ID using the SHA-1 which is then encrypted using the private key of the client to make a digital signature for him and subsequently a message digest. This technique allows sever to detect any data modification in the message sent by the client. The message digest is then encrypted using AES generated session to avoid any spoofing. The mobile application will encrypt the client's account ID by using the server's public key to retrieve the client's password. Finally both of the encrypted account ID and the encrypted message digest are sent to the bank server. [29]

3.1.7.4 Server protocol initialization

When the bank server starts up, the administrator has to login its password which is used to retrieve the server private key from sever key store. This step is done because all server private keys are saved in the key store and it could be signed only by using the administrator password.

3.1.7.5 SGP handshake server

When the user establishes a connection with server, the first message received by the server is the user public key .Following this, the server receives the full SGP message that will be immediately divided into encrypted message digest and encrypted account ID. By using the private key, the server retrieves the sent account ID and the client password .Therefore, if these steps fail to decrypt the message or the account ID is not available in the database, an error message will be sent to the client. Figure 11 shows unpacking and verifying the full SGP.

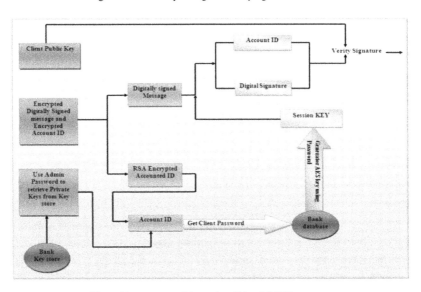

Figure 12: server unpacking and verifying full SGP message

The server generates the key by using retrieved password which is the same key to decrypt message digest. If the sever fails, then error message will be sent to the user. Otherwise, the server will split into two parts as explained above. To complete the handshake; the server will send a message to the client assuring him that a secure tunnel has been constructed. The client then should decode this message using the exchanged cipher suites.

It is to be mentioned here that the SGP record protocol achieves message confidentiality and message integrity. It is almost the same as the SGP handshake protocol; however the absence of the encrypted client account ID in the SGP message is the primary difference. [36]

3.1.7.6 Keys and certificates storage in the bank server.

The password which is encrypted can be saved in the MYSQL database, a key store file, via using the server's private key and can be unlocked only by a key generated by the administrator's password. It is to be mentioned here that each client has a uniquely encrypted server's public key using a PBE AES based on the client's password. No single session key is stored because session keys can be only generated during communication with the client. For immediately; the client generates a one –time512RSA key pair. Any change in the client's password necessitates updating the server public key which is to be sent to the server   over the air programming (OTA). [38]

In the SGP protocol, the SGP uses SHA-1 for the message hashing, 2048 RSA keys for the bank server and 128 bit AES keys. As for the client, RSA keys 512 bits, modified in a way to meet the low power of mobiles, are used.

3.1.7.7 Secure GPRS Protocol

The secure GPRS corresponds with the general security requirements again through the following:

1. Confidentiality. This protocol has been designed to take care of the banking security requirements. It ensures that the data or the information should be highly confidential in both bank and mobile application via using the AES encryption.

2. Integrity .In order to validate the integrity of the information (data), the protocol should be able to detect any modification or generation in the data from. Therefore, the bank server or mobile application should check the RSA digital signatures

3. Authentication .This prerequisite can be established by SGP as each mobile application is packed with the server certificate. This certificate contains the server public key which is used to authenticate the server. Simultaneously the server uses the client SGP certificate in order to authenticate the client.

4. Availability .To avoid replay the bank server detects decayed messages by checking timestamps for each message it receivers. This step provides great availability to the system and at the same time supports the security performances.

Two solutions had been provided in this project in order to solve the security issues in mobile banking and the third solution which concerning about the generation of the signal noise has been mentioned in the appendices part. So for further information please go to appendix A.

4. Result s of this project

The following table is a comparison between the prevailing mobile banking and our suggested solutions:

	existing solution			Suggestion solution		
	USSD and SMS	WIG and SMS	WAP	Secure SMS Message	expand the WAP	Secure the GPRS
Security of mobile banking	Uses Plaintext as string sent in, and the Authentication depends on the IMEI	Uses both the USSD and the SMS message transmitted in the plaintext	Stands on the WILS and does not have End to End encryption	Uses the AES symmetric key and one time password encryption to secure the SMS message. also it uses the SHA1 message with exclusive one time of password without repudiation	The GRPS Standard security, uses handshaking and full handshaking to make authentication with keys exchanges	Makes End to end security with the AES encryption and one time keys supporting the security. Also uses the RSA digital signatures and certification to make it more secure.
price for users	Using the USSD is free plus one message will be required	Using a Multiple message as requested	Usually depends on the quantity of the information that wants to be sent. The GPRS is normally cheaper than the SMS message	Only one message is used for each transaction	Depends on the quantity of the information or data. The GPRS is cheaper compared with SMS message	Depends on the quantity of information which is necessary to be sent. And the GPRS is cheap compared with the SMS message.
Price for Banking	Only one message needed for reply	Many SMS messages are required	Similar to the price for users	One SMS message is needed for reply	Similar to the price for users	Similar to the price for users

Kind and dependability of connection	The USSD is synchronous, also the user can get response directly from the sever of bank	The transaction is asynchronous as each one waits for a reply from the sever. Also the SMS message is stored in message buffer which is SMSC until the message reaches sever.	**Synchronous when the connection is lost and is then reconnected with the bank server.**	**Asynchronous. The SMS message will be stored in message buffer queue of sever until it reaches sever. This technique provides guarantees message delivery**	**Synchronou s**	**Synchronous**
The compatibility	Every mobile phone can provide the USSD and the SMS which can use this sever	Depends on the SIM card	Requires mobile phone to be the WAP capable and the GPRS, 3G or EDGE enabled	Depends on the improvement of the tools which can be used for implementation of the solution s	Requires mobile phone to be WAP capable and the GPRS, 3G enabled	Depends on the improvement and tools which can be used for implementation of the solution
user ability	Depends on the way users cooperates with mobile phones in sending SMS	The menu depends on the user interaction	Using mobile WAP to browse interface	Depends on the development and the tools that are going to be used	Using mobile WAP to browse interface	Depends on the development and the tools which are going to used

5. Conclusions & Future work

Needless to say here that this project has considered crucial issues of security which the mobile banking suffers from GSM and the GPRS security architecture has been approached from different angles, such as the security mechanisms in GSM network and how authentication can be processed throughout the whole network processes. Then, the SMS banking in Oman has been dealt with, the way it performs and the most crucial security problems in the SMS message. Furthermore, solutions have been suggested which can hopefully improve the performance security in mobile banking. Then, issues related to GPRS security and then some solutions have been proposed to solve these issues. Therefore, a simulation with the MATLAB has been carried out in an attempt to reveal the way a signal can be generated with and without channel noise to see the influence of these varied channels on the signal. In the suggested program, the OFDM technique has been used to generate the signal. The analysis of the OFDM proves to be useful with telecommunication which is mentioned in the Appendices part. The project has attempted to provide high security performance to the mobile banking, but unfortunately, no perfect security has been created so far, because everything depends on the user in feeding the system with information. Any failing in the system is definitely caused by wrong information (data) supplied by the user.

Undoubtedly, the future will open new vistas to the mobile banking through using and developing different programs such as C++ or Java. In addition, we hope that the security processes will be more complex and complicated, because it will protect the system against any spoofing and makes it easier to be used by customers. Furthermore, we aspire that a better database will be created by using the MYSQL to store and retrieve the information of the customer in a better and more secure way. Perhaps these aspirations could be the subject of the broader studies in the future.

References

1. Biryukov A., Shamir A., and Wagner D., (2001) "Real time cryptanalysis of A511 on a pc," in *Advances in Cryptology, proceedings of Fast Software Encryption* '00, vol. 1978, pp. 1-18, Springer-Verlag.

2. Herzberg A., (2003), *"Payments and Banking with Mobile Personal Devices,"* Communications of the ACM,Volume 46, Issue 5, pp. 53-58.

3. Dukic B. and Katic M., (2005),"m-order - payment model via SMS within the m-banking," *27th Int. Conference on Information Technology Interfaces*, 20-23 June, pp. 93-98.

4. Kasım B. and Ertaul L., (2000), "Evaluation of GSM Security", in Proc. of the 5th Symp. on Com. Networks (BAS 2000). (Accessed 22/6/2011)

5. Buchanan G., Farrant S., Jones M., Thimbleby H. Marsden G. and Pazzani M. *"Improving mobile internet usability,"* Proceedings of the tenth international conference on World Wide Web. ACM Press, New York, NY, 2001, 673-680. (Accessed 22/6/2011)

6. Cell phone Banking - How do I. (Website)https://www.fnb.co.za/personal/transact/accessyouraccounts/cellHowDoI.html (Accessed 10/7/2011)

7. Thanh D., "Security issues in mobile ecommerce," in Proc.of 11th International Workshop on Database and Expert Systems Applications, Greenwich, London, UK, pp. 412–425, Sep. 2000. (Accessed 19/6/2011)

8. D. Van Thanh, *"Security issues in mobileecommerce,"* in Proc.of 11th International Workshopon Database and Expert Systems Applications,Greenwich, London, UK, pp. 412–425, Sep. 2000.(Accessed 19/6/2011)

9. Margrave D., (2006) GSM Security and Encryption, Available link http://www.hackcanada.com/blackcrawl/cell/gsm/gsm-secur/gsm-secur.html (Accessed 23/7/2011)

10. Design and Implementation of Orthogonal Frequency Division Multiplexing (OFDM) Signaling http://cegt201.bradley.edu/projects/proj2001/ofdmabsh/OFDM_Design_Proposal.html (Accessed 5/2/2011)

11. ETSI, "Digital Cellular Telecommunications System Security Management", GSM 12.03 version 7.0.1 Release 2000. (Accessed 22/6/2011)

12. Scornavacca E., and Barnes S., "M-banking services: a strategic perspective", *International Journal of Mobile Communications*, 2004, pp. 51-66. (Accessed 25/6/2011)

13. Narendiran C., Rabara S. and Rajendran N., 2010 End-to-end encryption in GSM, mobile banking networks System, , http://ieeexplore.ieee.org, (Accessed 12/7/2011)

14. Elliott G., and Tang H, "The wireless mobile Internet: an international and historical comparison of the European and American Wireless Application Protocol (WAP) and the Japanese iModeTM service,"Internation Journal of Information Technology and Management, 3, Nos. 2/3/4, pp. 268-281, 2010. (Accessed 21/6/2011)

15. Koien G, "An introduction to Access Security in UMTS", IEEE Wireless Communications, Vol.11, No 1, Feb 2004. (Accessed 23/6/2011)

16. Brookson G., GSM (and PCN) Security and Encryption , 1994(Accessed 15/7/2011

17. Hadson, Mobile Messaging Technologies And Services: SMS, EMS and MMS, 2nd ed., John Wiley & Sons Ltd., 2008. (Accessed 8/6/2011)

18. Herzberg A., 2003, Payments and banking with mobile personal devices. *Communications of the ACM* .Volume 46, Issue 5 (May 2003) Wireless networking security Pages: 53 58 ISSN: 0001-0782 (Accessed 2/7/2011)

19. Harb, and H. Farahat, "Secure SMS Pay: Secure SMS Mobile Payment Model," in proc. Automotive Mechanics Conference, pp. 110-115, 2008. (Accessed 6/6/2011)

20. Golic J., "Cryptanalysis of alleged AS stream cipher," in *Advances in Cryptology,* vol. 1233 of *LNCS,* pp. 239-255, Springer Verlag. (Accessed 15/6/2011)

21. Lauter K., "The Advantages of Elliptic Curve Cryptography for Wireless Security," *IEEE Wireless Communications,* vol. 11, no. 1, pp. 62-67, February 2004. (Accessed 23/6/2011)

22. Lord, S. 2003. Trouble at Telco: When GSM Goes Bad, *Network Security.* Issue 1, Page10 -12. (Accessed 5/7/2011)

23. Li Wei, Wu Qinghua, Liao Weiguo, WPKI Mobile Banking Security Technology Model Research. The Journal of Hubei Industrial Institution. 2004 (19). (Accessed 14/6/2011)

24. Toorani M., "SSMS - A Secure SMS Messaging Protocol for the M-Payment Systems," in proc. IEEE Symp. Computers Communications (ISCC), pp.700-705, 2008. (Accessed 10/6/2011)

25. Shirali-Shahreza M., "Stealth Steganography in SMS," Proceedings of the Third IEEE and IFIP Int. Conf. on Wireless and Optical Communications Networks, 2010. (Accessed 12/6/2011)

26. Briceno M., "A pedagogical implementation of the gsm A511 and A512"voice privacy "encryption algorithms." http://cryptome.org/gsma512.htm, 1999. (Accessed 15/6/2011)

27. Mobile Banking: Security and Fraud Issues http://www.mobilephonesandsafety.co.uk/mobile-banking-security-fraud-issues.html (Accessed 3/7/2011)

28. Mobile banking http://en.wikipedia.org/wiki/Mobile_banking (Accessed 1/8/2011)

29. Mobile security in mobile banking http://technology.cgap.org/2008/04/03/mobile-security-in-mobile-banking/(Accessed 4/8/2011)

30. Mobile banking http://www.barclays.co.uk/MobileBanking/Mobilebanking/P1242561069586 (Accessed 7/8/2011)

31. Toorani M. & Ashar A., (2010) Solutions to the GSM Security Weaknesses http://ieeexplore.iee.org.(Accessed 14/7/2011)

32. Kahzadi N., Edalat E. and Dehgan-Dehnavi M., "Commerce and M-Banking in World and Iran," *Proceedings of the Third National Conference on E-Commerce*, Tehran, Iran, 31 May-1 June, 2005, pp. 306-329 (In Persian). (Accessed 6/2/2011)

33. Mallat N. and Rossi M. and Tuunainen M, "Mobile banking services," Communications of the ACM, Volume, 47, Issue 5, pp. 42-46,2004. (Accessed 18/6/2011)

34. NBO upgrades its SMS banking services http://www.gulfbase.com/site/news/NBO-upgrades-its-SMS-banking-services_109854.aspx (Accessed 26/7/2011)

35.Overview of GSM and GSM Security Author: Tuan Huynh and Hoang Nguyen of Oregon State University Date: June 6, 2003(Accessed 25/7/2011)

36. Ekdahl P. and Johansson T., "Another attack on AS11 ,"*Transactions on Information Theory*, vol. 49, pp. 284-289, 2003 (Accessed 17/6/2011)

37.Soni P., "Mobile Payment Between Banks Using SMS," Proc. of IEEE, vol. 98, no. 6, pp. 903-905, June, 2010. (Accessed 6/2/2011)

38.Tiwari R. and Buse, The Mobile Commerce Prospects:"A Strategic Analysis of Opportunities in the Banking Sector, Hamburg University Press, Hamburg, 2008. (Accessed 8/6/2011)

39.Kungpisdan S., Srinivasan B., and P. D. Le, "*A secure account-based mobile payment protocol,*" presented at Information Technology: Coding and Computing, 2004. Proceedings, ITCC 2004. International Conference on Volume (Accessed 21/7/2011)

40."Security in the GSM System" By Jeremy Quirke (Accessed 21/7/2011)

41.Misra S., and Wickamasinghe N., "*Security of a mobile transaction: A trust model,*" Electronic Commerce Research 4 (4) (2004) 359–372. (Accessed 21/6/2011)

42. Lee S. and SeungBae Park: Mobile Password System for Enhancing Usability- Guaranteed Security in Mobile Phone Banking. Computer Science, 2005, pp. 66-74, (Accessed 14/6/2011)

43. Petrovic S. and Fuster-Sabater A., "Cryptanalysis of the A512 algorithm." Cryptology ePrint Archive, Report 2001052, http://eprint.iacr.org, 2000. (Accessed 18/6/2011)

44. Laukkanen T., "Comparing consumer value creation in Internet and mobile banking," International *Conference on Mobile Business (ICMB 2005)*, 11-13 July, 2005, pp. 655- 658. (Accessed 10/6/2011)

45. VLSI implementation of OFDM modem http://www.design-reuse.com/articles/10358/vlsi-implementation-of-ofdm-modem.html (Accessed 12/7/2011)

46. Vergados D.D., Liveris A.D., Anagnostopoulos C., Lykou C. and Stassinopoulos G.I. ,New Generation Features for GSM Systems 2010 http://ieeexplore.ieee.org.ezproxy.mdx.ac.uk/stamp/stamp.jsp?tp=&arnumber=905925 (Accessed 2/7/2011)

47. Chou Y., Lee C. and Chung J., "Understanding m-commerce payment systems through the analytic hierarchy process," Journal of Business Research, 2010(Accessed 2/7/2011)

48. Yong Z. and ChangZhen X., The Research on AES and ECC Integration Data Encryption Technology. Computer Security, 2010 (Accessed 14/6/2011)

Appendix A:
Practical section:

This section uses MATLAB software to design a program for mobile banking in an attempt to demonstrate the mechanism by which a signal can be generated from wireless telecommunication. OFDM, Orthogonal Frequency Division Multiplex will be therefore used. This modulation is being widely used nowadays in telecommunication of both wired and wireless connection. In the OFDM modulation scheme, a large number of sub channels are used to transmit a digital data rate sub-channels in which each modulates a single carrier. Sub-carriers are orthogonal because they are separated by the reciprocal of the sub-channel data rate. It will provide a high spectral efficiency. Using the OFDM has reduced the multipath interference which occurs in the receiver side. Therefore, simulation for two programs will be presented using MATLAB .The first program generates signal, transmits and receives OFDM signal whereas the other will use channel noise. These programs will explain the generation of signal either with or without channel noise or using OFDML modulation over the mobile phone network. Furthermore, a low pass filter has been used in the receiver side in order to remove any noise might occur with users when they are sending their personal information to the bank via the SMS.

Figure 13 shows the OFDM system block diagram which is going to be designed .This OFDM could be summarized in the following steps: [45]

6. The incoming serial data has high data rate R encoded by a complex modulation format.
7. The encoded data stream will be converted from the serial to the parallel to provide N of complex data sub channels with a data rate R/N to every channel. In addition, the sequence of the N parallel complex data sub channels will move from frequency domain to time domain through inverted fast Fourier transform IFFT process to generate the OFDM signal.
8. The OFDM converts the signal to serial data transmission. Moreover, the orthogonality between the different subs carriers is made effective via ISI inter symbol interference which can cause multipath propagation to the radio channel. By converting the signal to analogue with low pass filtered for the radio frequency, the problem is solved.
9. The OFDM receiver fundamentally inverses the OFDM transmitter. Therefore, the incoming signal is mixed down to base band filtering a converter to digital words. The FFT reviver extracts the phase and amplitude for each sub carrier, provided by the block receiver

samples, to give a successfully receiver signal demodulate. The FFT should start from the beginning of one of the OFDM data block as shown in figure 13.

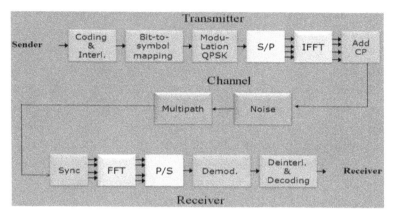

Figure 13: OFDM system block diagram

In the first program, no channel noise has been used so as to examine the OFDM signal without any affectation. Figure 14 shows the signal that wants to be transmitted. Random numbers of signals have been chosen to see if the signal will change or not. These parameters have been used:

```
M = 4;
no_of_data_points = 64;
block_size = 8;
cp_len = ceil(0.1*block_size);
no_of_ifft_points = block_size;
no_of_fft_points = block_size;
```

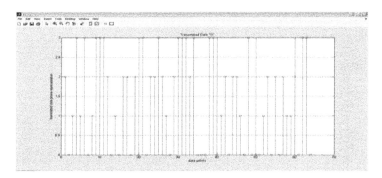

Figure 14: Transmitter Data

The following code has been used to set up the signal

```
data_source = randsrc(1, no_of_data_points, 0:M-1);
figure(1) stem(data_source); grid on; xlabel('data points'); ylabel('transmitted data phase representation')
title('Transmitted Data "O"')
```

Figure 15 shows the modulated that needs to be transmitted. There are 4 points in this signal.

This means that 4 QPSK modulated have been used to detect if there are any changes in the sig-

nal.

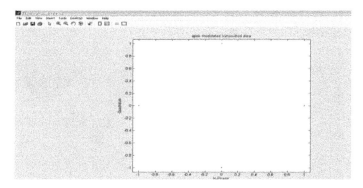

Figure 15: QPSK modulated (transmitter Data)

QPSK (Quadrature Phase Shift Keying) is a phase modulation algorithm; a version of frequency modulation where the phase of the carrier wave is modulated to encode bits of digital information in each phase change. In this case, QPSK refers to the use of phased shift keying and is accomplished by the use of 4 discrete states. Its significance arise the need to see how the performance of the signal varies before and after transmitting the data.

This setting has been used to show this modulated

```
qpsk_modulated_data = pskmod(data_source, M);
scatterplot(qpsk_modulated_data);title('qpsk modulated transmitted data')
```

Furthermore, the root concepts of QPSK and 4 QAM are different; the resulting modulated radio waves are exactly the same. QPSK uses four points on the constellation diagram, equispaced around a circle. With four phases, QPSK can encode two per symbol shown in the diagram with gray coding (each adjacent symbol only differs by one bit) to minimize the bit error rate (BER). In other words, the QPSK can be used either to double the data rate in comparison with a BPSK system while maintaining the same bandwidth of the signal or to maintain the data of BPSK but with the necessary bandwidth. In this cause BER of QPSK is exactly the same as the BER of BPSK. Usually, confusion arises when considering QPSK; radio communication channels are allocated by agencies such as the federal communication commission that gives a prescribed maximum bandwidth. In this respect, the advantage of QPSK over BPSK becomes quite evident as QPSK transmits twice the data rate in given bandwidth compared to BPSK. QPSK transmitters and receivers are more complicated than those of BPSK. However, with modern electronics technology, the penalty in cost becomes very moderate. But as with BPSK, there are phase ambiguity problems at the receiving end and therefore varying encoded QPSK is often used for more practical functionality.

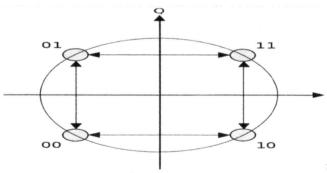

Figure 16: 4 QPSK

Figure 17, shows the OFDM signal implementation without any channel noise.

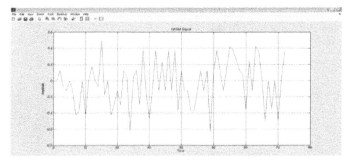

Figure 17: OFDM signal

The OFDM is adopted as the modulation method because all the new technologies being used and developed are subject to continuous development and advancement these days. It is probably the most efficient method discovered so far. It mitigates the sever problem of multipath propagation that causes huge data error and loss of signal in the microwave and UHF spectrum. In addition, OFDM is based on the concept of frequency division multiplexing (FDD) in which the medium could be radio spectrum, coax cable, twisted pair, or fiber optic cable. Each data stream is modulated into multiple adjacent carriers within the bandwidth of the medium, and is transmitted simultaneously. The serial digital data stream to be transmitted is subsequently split into multiple slower data streams, and each is modulated into a separate carrier in specific spectrum which is called subcarrier. This modulation is used with digital data, but the quadrature

Phase shift keying (QPSK) is the most common, and this modulation is used in this simulation. In addition, using FFT is a good way to separate the carriers of an OFDM signal. Furthermore, the used IFFT just reverses the FFT process. All the individual carriers with modulation are in digital form and are then subjected to an IFFT mathematical process creating a single composite signal that can be transmitted. The FFT revives all sorts of signals to recreate the original data stream. Figure 18 shows FFT and IFFT work processes.

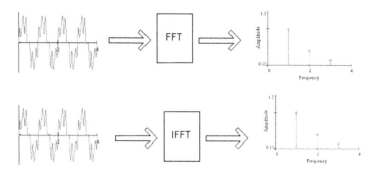

Figure 18: Time domain signal comes out as spectrum out of a FFT and IFFT

The advantages of using OFDM are many but the most important one id its spectral efficiency, also called bandwidth efficiency, this means that you can transmit more data but faster inn way and bandwidth even in the existence of noise. The measure of spectral efficiency is bits per second Hertz, or bps/Hz. Different modulation methods can give user maximum and different data rates for given bit error rate (BER) and noise level. Simple digital modulation methods like amplitude shift keying (ASK) and frequency shift keying (FSK) are fair and simple. BPSK and QPSK are much better when used with OFDM.

The code which is used to the implementation of this signal is show below

```
ofdm_signal = reshape(ifft_data, 1, len_ofdm_data);
figure(3)plot(real(ofdm_signal)); xlabel('Time'); ylabel('Amplitude');
title('OFDM Signal');grid on;
```

Figure 19 shows the 4 QPSK modulations in the received data. The figure clearly displays that there are not any noise or interference in the modulation, because in figure 15, 4 point s are set at zero.

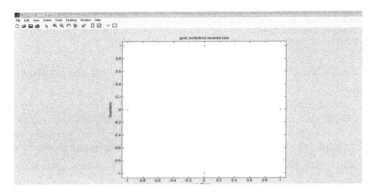

Figure 19: QPSK modulation (received Data)

Set up QPSK modulation at the receiver data uses this function

qpsk_demodulated_data = pskdemod (recvd_serial_data, M);
scatterplot(qpsk_modulated_data);title('qpsk modulated received data'

Figure 20 shows the received data .A compare figure 14 which is the transmitter data with this figure; showed that the signal is the same, without any change or noise affect. The signal does not change because no tool was used to add the noise on this signal. The reasons behind this, is examine the signal without any noise interference.

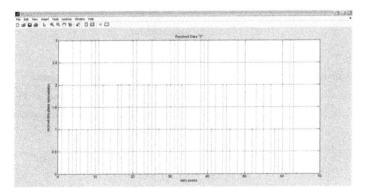

Figure 20: Receiver data

To set up the received data, this code has been used.

```
figure(5)stem(qpsk_demodulated_data,'rx');
grid on;xlabel('data points');ylabel('received data phase representation');title('Received
Data "X"')
```

In the second program, channel noise will be used. , we will note that there is a lot of change in the signal.

By following the same code that was used in the first program, the channel noise will be added

Figure 21 displays the signal that needs to be transmitted, here; a random number was used to transmit data by any user. The reason behind this data is to test how the signal received when a channel noise is added.

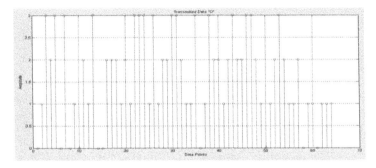

Figure 21: transmitter data

Figure 22 shows the received data modulation which has many changes because of the noise effect. Also, it describes the 4 QPSK which was used in zero regions, but when the channel noise was added in this part, it affected the quality of the signal and caused interference with the signal.

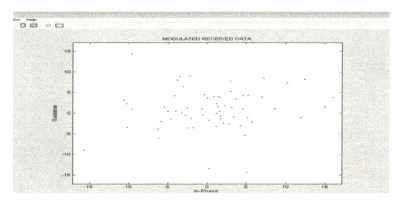

Figure 22: QPSK receiver data

This code is used to display modulation with channel effect.

```
qpsk_demodulated_data = pskdemod(recvd_serial_data,M);
scatterplot(recvd_serial_data);title('MODULATED RECEIVED DATA');
figure(5)stem(qpsk_demodulated_data,'rx');
grid on;xlabel('Data Points');ylabel('Amplitude');title('Received Data "X"')
```

Figure 23 shows the OFDM after using the channel Noise. A big change occurs in this signal. Comparing figure 17 with this figure shows that the signal in figure 17 starts from zero and ends with 2.4, whereas the signal in 20 starts from 0.4 and ends at -0.4 because of the effect of the channel noise.

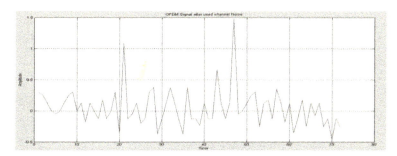

Figure 23: OFDM after channel Noise

Figure 24 shows the received data. Going back to figure 21 exhibits that there is a difference between sending which starts from one and ends at one and the received signal which starts from 2 and ends in 3 level because of the effect of noise.

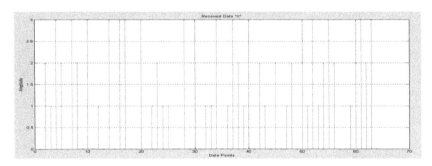

Figure 24: received data

To set up this signal, the following code is used;

```
scatterplot(recvd_serial_data);title('MODULATED RECEIVED DATA');
figure(5)stem(qpsk_demodulated_data,'rx');
grid on;xlabel('Data Points');ylabel('Amplitude');title('Received Data "X"')
```

The low pass filter is concerning about connecting both of the signal resistor and the signal capacitor in series connection. Therefore, this type of filters arranges the resistor and the capacitor to be on the input side. On the other hand the output signal is taken across the capacitor only as the following circuit shows:

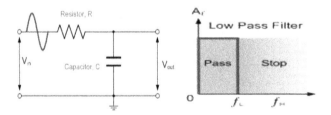

Figure 25: Low Pass Filter Circuit   Ideal Filter Response Curves

In the low pass filter the frequency increases from 100 to 10K Hz, on the other hand the output voltage decreases from 9.9 to 0.718v. Therefore, by plotting both of the output and the input frequency, the frequency response curve of the low pass filter can be given by the following graph:

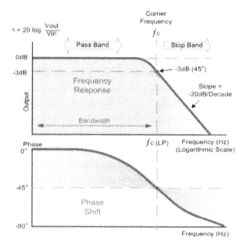

Figure 26: Frequency signal Response use Low pass Filter

The above plots show that the frequency response of the filter is nearly flat at the low frequency. So at this stage all the input signals have the ability to pass directly to the output until it reaches the cut off frequency point (fc). That is happening because the reactance of the cut off frequency point is very high at the low frequencies and it blocks any current flow to go through the capacitor. After the fc point the circuit decreases as what the above graph shows that the slop = (-20db/ Decade). The decreases will continue until the reactance of the capacitor becomes very low and after a while the output terminals resulting will reach the zero output.

However, by using the above technique (LPF) it will be easier to reduce the noise channel affection which might occur when the user sends a message to the bank server.

Appendix B:

MATLAB software is used in this project, because it is more flexible and easy to be used. In addition, it provides great graphic diagrams and is suitable to test a signal performance with or without noise. Therefore, the OFDM module has been used to generate the signal.

Program 1: Generating a signal with no noise effect

```
Clear all
clc
close
M = 4;
no_of_data_points = 64;
block_size = 8;
cp_len = ceil(0.1*block_size);
no_of_ifft_points = block_size;
no_of_fft_points = block_size;
data_source = randsrc(1, no_of_data_points, 0:M-1);
figure(1)
stem(data_source); grid on; xlabel('data points'); ylabel('transmitted data phase representation')
title('Transmitted Data "O"')
qpsk_modulated_data = pskmod(data_source, M);
scatterplot(qpsk_modulated_data);title('qpsk modulated transmitted data')
num_cols=length(qpsk_modulated_data)/block_size;
data_matrix = reshape(qpsk_modulated_data, block_size, num_cols);
cp_start = block_size-cp_len;
cp_end = block_size;
for i=1:num_cols,
ifft_data_matrix(:,i) = ifft((data_matrix(:,i)),no_of_ifft_points);
for j=1:cp_len,
actual_cp(j,i) = ifft_data_matrix(j+cp_start,i);
end
ifft_data(:,i) = vertcat(actual_cp(:,i),ifft_data_matrix(:,i));
end
[rows_ifft_data cols_ifft_data]=size(ifft_data);
len_ofdm_data = rows_ifft_data*cols_ifft_data;
ofdm_signal = reshape(ifft_data, 1, len_ofdm_data);
figure(3)
plot(real(ofdm_signal)); xlabel('Time'); ylabel('Amplitude');
title('OFDM Signal');grid on;
recvd_signal = ofdm_signal;
recvd_signal_matrix = reshape(recvd_signal,rows_ifft_data,cols_ifft_data);
recvd_signal_matrix(1:cp_len,:)=[];
for i=1:cols_ifft_data,
fft_data_matrix(:,i) = fft(recvd_signal_matrix(:,i),no_of_fft_points);
end
recvd_serial_data = reshape(fft_data_matrix, 1,(block_size*num_cols));
qpsk_demodulated_data = pskdemod(recvd_serial_data,M);
scatterplot(qpsk_modulated_data);title('qpsk modulated received data')
figure(5)
stem(qpsk_demodulated_data,'rx');
grid on;xlabel('data points');ylabel('received data phase representation');title('Received Data "X"'
```

Program 2: Generating a signal with noise effect

```
clear all
clc
close
M = 4;
no_of_data_points = 64;
block_size = 8;
cp_len = ceil(0.1*block_size);
no_of_ifft_points = block_size;
no_of_fft_points = block_size;
data_source = randsrc(1, no_of_data_points, 0:M-1);
figure(1)
stem(data_source); grid on; xlabel('Data Points'); ylabel('Amplitude')
title('Transmitted Data "O"')
qpsk_modulated_data = pskmod(data_source, M);
scatterplot(qpsk_modulated_data); title('MODULATED TRANSMITTED DATA');
num_cols = length(qpsk_modulated_data)/block_size;
data_matrix = reshape(qpsk_modulated_data, block_size, num_cols);
cp_start = block_size-cp_len;
cp_end = block_size;
for i=1:num_cols,
ifft_data_matrix(:,i) = ifft((data_matrix(:,i)),no_of_ifft_points);
for j=1:cp_len,actual_cp(j,i) = ifft_data_matrix(j+cp_start,i);
end ifft_data(:,i) = vertcat(actual_cp(:,i),ifft_data_matrix(:,i));
end[rows_ifft_data cols_ifft_data]=size(ifft_data);
len_ofdm_data = rows_ifft_data*cols_ifft_data;
ofdm_signal = reshape(ifft_data, 1, len_ofdm_data);
figure(3)plot(real(ofdm_signal)); xlabel('Time'); ylabel('Amplitude');
title('OFDM Signal');grid on;
noise = randn(1,len_ofdm_data) + sqrt(-1)*randn(1,len_ofdm_data);
avg=0.4;for i=1:length(ofdm_signal)
if ofdm_signal(i) > avg
ofdm_signal(i) = ofdm_signal(i)+noise(i);
endif ofdm_signal(i) < -avg ofdm_signal(i) = ofdm_signal(i)+noise(i);
endfigure(4)plot(real(ofdm_signal)); xlabel('Time'); ylabel('Amplitude');
title('OFDM Signal after used channel Noise');grid on;
channel = randn(1,block_size) + sqrt(-1)*randn(1,block_size);
after_channel = filter(channel, 1, ofdm_signal);
awgn_noise = awgn(zeros(1,length(after_channel)),0);
recvd_signal = awgn_noise+after_channel;
recvd_signal_matrix = reshape(recvd_signal,rows_ifft_data,cols_ifft_data);
recvd_signal_matrix(1:cp_len,:)=[];
for i=1:cols_ifft_data,fft_data_matrix(:,i) = fft(recvd_signal_matrix(:,i),no_of_fft_points);
endrecvd_serial_data = reshape(fft_data_matrix, 1,(block_size*num_cols));
scatterplot(recvd_serial_data);title('MODULATED RECEIVED DATA');
qpsk_demodulated_data = pskdemod(recvd_serial_data, M);
scatterplot(recvd_serial_data);title('MODULATED RECEIVED DATA');
figure(5)stem(qpsk_demodulated_data,'rx');
grid on;xlabel('Data Points');ylabel('Amplitude');title('Received Data "X"')
```

It is important to mention again that this program was designed by using the MATLB. The main purpose was to comprehend fully the mechanism by which a signal is generated via the OFDM with or without channel noise. The suggested program enables us to validate the assumed signal and how it can affect the quality and performance of the signal upon receiving.

www.ingramcontent.com/pod-product-compliance
Lightning Source LLC
LaVergne TN
LVHW042349060326
832902LV00006B/485